Mitteilungen aus dem Institut für angewandte Mathematik
AN DER EIDGENÖSSISCHEN TECHNISCHEN HOCHSCHULE, ZÜRICH
HERAUSGEGEBEN VON PROF. DR. E. STIEFEL

Nr. 4

An Oscillation Theorem for Algebraic Eigenvalue Problems and its Applications

by

Frank W. Sinden

Springer Basel AG
1954

ISBN 978-3-0348-4075-0   ISBN 978-3-0348-4149-8 (eBook)
DOI 10.1007/978-3-0348-4149-8

Nachdruck verboten. Alle Rechte, insbesondere
das der Übersetzung in fremde Sprachen und der Reproduktion
auf photostatischem Wege oder durch Mikrofilm, vorbehalten
Copyright 1954 by Springer Basel AG
Originally published by Verlag Birkhäuser AG., Basel in 1954.
Unveränderter Nachdruck 1967

CONTENTS

                                                                            page

Introduction   7

Part I

1. Interior and border vectors   9

2. An algebraic oscillation theorem   11

3. Proof of Theorem 1   13

4. Checkerboard transformation and proof that the corollary is a consequence of the theorem   19

Part II

5. Mean value operator   23

6. Difference operator   31

7. Extensions   39

8. Applications   41

Appendix

9. Vibrating rod with intermediate supports.   45

German Summary   51

References   57

Curriculum vitae   59

Introduction

Oscillation theorems of the Sturm-Liouville type have been proved for a wide variety of one dimensional eigenvalue problems including many of higher order and with relatively complicated boundary conditions. Today, with the increasing importance of numerical methods, such theorems are of interest not only for the exact analytical eigenvalue problems, but for the approximating algebraic ones as well. In the first part of the present paper an oscillation theorem is proved for a certain class of matrix operators, and in the second part some examples and applications in the field of practical mathematics are given.

The theorem of Part I is not based on any direct analogy between the algebraic problem and a differential problem (e.g. via difference equations), but is related rather to the comprehensive theory of "oscillatory" operators which was originated by O. Kellogg and developed by M. Krejn.[1] This theory is principally concerned with integral equations whose kernels are of the so-called "totally non-negative" or "Kellogg" type. The eigenfunctions of such kernels display oscillatory properties of the familiar Sturm-Liouville type. As a matter of fact the classical Sturm-Liouville oscillation theorem as well as many of its important extensions are included as special cases. In the following we shall employ in place of the property "totally non negative" the roughly equivalent*) and more graphic property "variation-diminishing".

---

*) e.g. they are equivalent when the matrix is positive definite.

Part I

1. INTERIOR AND BORDER VECTORS.

It will be convenient to use vector notation even though we are not concerned with vectors in the sense of entities which are invariant under coordinate transformations. We shall therefore use the word "vector" simply to mean a matrix of one row and n columns (or n rows and one column as the case may be). The components of a vector are to be thought of as representing a table of a continuous function. Instead of the number of nodes of an eigenfunction, we consider the number of variations of sign in the sequence of components of an eigenvector.

Definition. The sequence $Q = (q_1, q_2, \ldots, q_n)$ is said to have a <u>variation of sign</u> between the elements q_{i-m} and q_i if

$$q_{i-m} \, q_i < 0$$

and if $q_\nu = 0$ for all ν such that $i-m < \nu < i$, where

$$0 < m < i.$$

Definition. The <u>variation number</u> of a vector φ, denoted by $V(\varphi)$, is the number of variations of sign in the sequence of components of φ.

Lemma 1. For every vector φ there exists an $\varepsilon > 0$ such that

$$V(\varphi + \vartheta) \geq V(\varphi)$$

for every vector ϑ satisfying the relation $(\vartheta, \vartheta) < \varepsilon.$ *)

In other words, an infinitesimal change in a vector may increase but can never decrease its variation number.

The one-sidedness of this proposition reflects the one-sided definition of "variation number". The sequence (1, 0, 1), for example, whose variation number is zero, could with equal justification be assigned the variation number two.

---

*) (ϑ, ϑ) stands for the scalar product of ϑ with itself.

Definition. If there exists an $\varepsilon > 0$ such that
$$V(\tilde{e} + \vartheta) = V(\tilde{e})$$
for every ϑ such that $(\vartheta, \vartheta) < \varepsilon$, then $\tilde{e}$ will be said to be an <u>interior vector</u>. Otherwise $\tilde{e}$ will be said to be a <u>border vector</u>. *)

Lemma 2. If $\tilde{v}_k$ is any sequence of vectors such that **)
$$\lim \tilde{v}_k = \tilde{m}$$
where $\tilde{m}$ is an <u>interior</u> vector, then there exists a number N such that for all $k > N$, $V(\tilde{v}_k) = V(\tilde{m})$.

Lemma 3. $\tilde{e} = (x_1, ..., x_n)$ is an interior vector if and only if

$x_1 \neq 0$

$x_i = 0$ only when $x_{i-1} x_{i+1} < 0$, $\quad i = 2, ..., n-1$,

$x_n \neq 0$

Lemma 4. For every pair of linearly independent vectors $\tilde{e} = (x_1, ..., x_n)$ and $\tilde{y} = (y_1, ..., y_n)$ such that $V(\tilde{e}) = V(\tilde{y}) = V$, there exists a linear combination $\tilde{z} = a\tilde{e} + b\tilde{y}$ such that $V(\tilde{z}) \neq V$.

---

*) All vectors in the same octant of 3-dimensional space have the same sign sequence. With each octant, therefore, there is associated a variation number. A border vector is a vector which lies in a coordinate plane separating two octants with different variation numbers. A border vector always has the variation number of that neigboring octant whose variation number is lowest.

**) Componentwise convergence is understood.

Proof. Let k be the smallest value of i for which $x_i^2 + y_i^2 \neq 0$.
For the sake of definiteness let $x_k \neq 0$. Choose a and b so that $z_k = 0$.

Let
$$\mathbf{z}' = \mathbf{z} + \varepsilon \mathbf{x}$$

In particular
$$z_k' = \varepsilon x_k \neq 0$$

If $|\varepsilon|$ is sufficiently small (Lemma 1), then the sign of ε can always be chosen so that $V(\mathbf{z}') > V(\mathbf{z})$. Therefore either $V(\mathbf{z}) \neq V$ or $V(\mathbf{z}') \neq V$.

2. AN ALGEBRAIC OSCILLATION THEOREM. *)

<u>Definition.</u> The real **) matrix A of the linear transformation $A\mathbf{x} = \mathbf{y}$ is said to be <u>variation-increasing</u> if $V(A\mathbf{x}) \geq V(\mathbf{x})$ for every vector $\mathbf{x}$, and <u>variation-diminishing</u> if $V(A\mathbf{x}) \leq V(\mathbf{x})$ for every vector $\mathbf{x}$.

---

*) The theorem of this section may be deduced directly from a theorem of Gantmakher and Krejn (ref. 2, théorème 14) with the aid of a theorem of Schoenberg (ref. 3, Satz 2). The proof given in the present paper is elementary in character; in particular it is free of determinant theory.

**) All quantities in this paper are assumed to be real, although this is generally not explicitly stated.

Theorem 1. Consider the eigenvalue problem

(1) $$\mathcal{O}\tilde{\varkappa} = \lambda \mathcal{S}\tilde{\varkappa},$$

where $\mathcal{O}$ is a symmetric, positive definite matrix, and $\mathcal{S}$ a positive diagonal matrix.*). If $\mathcal{O}$ is variation-diminishing, and if all its codiagonal elements $a_{i-1,i}$, $a_{i+1,i}$ are different from zero, then all the eigenvalues of problem (1) are simple and positive:

$$\lambda_1 > \lambda_2 > ... > \lambda_n > 0.$$

If $\tilde{M}_k$ is the eigenvector which corresponds to λ_k, then

$$V(\tilde{M}_k) = k - 1.$$

More generally, if $\mathcal{MO}$ is any linear combination of eigenvectors:

$$\mathcal{MO} = \sum_{i=r}^{s} c_i \tilde{M}_i \quad \text{where} \quad 1 \leq r \leq s \leq n,$$

then
$$r - 1 \leq V(\mathcal{MO}) \leq s - 1.$$

An eigensolution **) having these properties will be called <u>inversely oscillatory</u>.

---

*) i.e. the elements of $\mathcal{S}$ satisfy the relations:

$$d_{ij} \begin{cases} = 0 & \text{for } i \neq j \\ > 0 & \text{for } i = j \end{cases}$$

**) By "eigensolution" is meant the totality of eigenvalues and eigenvectors.

Corollary. If $\mathcal{O}$ is variation-increasing, and all other conditions of Theorem 1 are satisfied, then

$$V(\tilde{\mathcal{M}}_k) = n - k.$$

More generally, if

$$\mathcal{M} = \sum_{i=r}^{s} c_i \tilde{\mathcal{M}}_i \quad \text{where} \quad 1 \leq r \leq s \leq n$$

then
$$n - r \geq V(\mathcal{M}) \geq n - s$$

Such an eigensolution will be called <u>oscillatory</u>.

3. PROOF OF THEOREM 1.

<u>Lemma 5.</u> All of the eigenvectors of problem (1) are interior vectors, i.e. if $\tilde{\mathcal{M}}$ is an eigenvector, then $u_1 \neq 0$, $u_n \neq 0$, and $u_i = 0$ implies $u_{i-1} u_{i+1} < 0$ for $1 < i < n$.

Proof. This lemma depends on the fact that the codiagonal elements of $\mathcal{O}$ are different from zero.

Let $\mathcal{M} = \mathcal{O}\tilde{\mathcal{M}} = \lambda \mathcal{D} \tilde{\mathcal{M}}$. Since $\mathcal{D}$ is a positive diagonal matrix, the sign sequence of $\mathcal{M}$ is exactly the same as that of $\tilde{\mathcal{M}}$.

Suppose $u_1 = 0$. From this assumption we shall derive a contradiction. Let u_k be the first non-vanishing component of $\tilde{\mathcal{M}}$. We now form a new vector $\tilde{\mathcal{M}}'$ by adding a small quantity δ, whose size and sign are yet to be determined, to the component u_k:

$$u'_k = u_k + \delta$$
$$u'_i = u_i \quad \text{for} \quad i \neq k.$$

Let the transform of $\tilde{\mathcal{M}}'$ be $\mathcal{M}' = \mathcal{O}\tilde{\mathcal{M}}'$. We shall now show that by suitable choice of δ it is always possible to satisfy the relation $V(\mathcal{M}') > V(\tilde{\mathcal{M}}')$, which contradicts the assumption that $\mathcal{O}$ be variation-diminishing.

Let $|\delta|$ be so small that

(1) $u_k \cdot u'_k > 0$ (This implies $V(\tilde{\mathcal{M}}) = V(\tilde{\mathcal{M}}')$.)

(2) $v_i \cdot v'_i > 0$ for all i such that $v_i \neq 0$. *)

As a matter of fact let us suppose that $|\delta|$ is so small that these conditions hold regardless of the sign of δ.

$\mathcal{A}\mathcal{O}'$ may be written

$$\mathcal{A}\mathcal{O}' = \mathcal{O}\mathcal{L}\tilde{\mathcal{M}}' = \mathcal{A}\mathcal{O} + \delta \alpha_k,$$

where $\alpha_k = (a_{1k}, \ldots, a_{nk})$ is the vector whose components are the elements of the k-th column of $\mathcal{O}\mathcal{L}$. In particular

$$v'_{k-1} = \delta a_{k-1,k}$$

which is different from zero because $a_{k-1,k}$ is a codiagonal element. By condition (2) v'_k is also different from zero **).

If now we choose the sign of δ so that

$$v'_{k-1} v'_k < 0$$

then by virtue of Lemma 1 we are led to the impossible relation

$$V(\mathcal{A}\mathcal{O}') > V(\mathcal{A}\mathcal{O}) = V(\tilde{\mathcal{M}}) = V(\tilde{\mathcal{M}}').$$

Therefore our assumption that $u_1 = 0$ must be false.
Similarly for u_n.

It remains to show that $u_i = 0$ implies $u_{i-1} u_{i+1} < 0$ for $i = 2, \ldots, n-1$.
The principle of the proof is the same as before.

---

*) It follows from (2) that $V(\mathcal{A}\mathcal{O}') \geq V(\mathcal{A}\mathcal{O})$. (Lemma 1.)

**) The reader will recall that u_k and v_k were assumed to be the first non-vanishing elements of $\tilde{\mathcal{M}}$ and $\mathcal{A}\mathcal{O}$ respectively.

Suppose that for some k

$$u_{k+\nu} = 0 \qquad \nu = 1, \ldots, \ell-1.$$

Case 1: $\qquad u_k u_{k+\ell} > 0.$

As before we form the vector $\check{M}'$ by adding a small quantity δ to the component u_k. If $|\delta|$ is sufficiently small, then it is always possible to choose the sign of δ so that

$$V(\Lambda D') > V(\check{M}'),$$

which is impossible. (See Fig. 1.)

Case 2: $\qquad u_k u_{k+\ell} < 0, \qquad \ell-1 \geq 2.$

In this case we need two small quantities, δ_1 and δ_2. Let

$$u'_k = u_k + \delta_1,$$

and for some $j \neq k$, which is to be determined later, let

$$u'_j = u_j + \delta_2.$$

Let us assume for the sake of definiteness that $u_k > 0$ and $u_{k+\ell} < 0$. By suitable choice of δ_1 and δ_2 we wish to satisfy the inequalities

$$a_{k+1,k}\delta_1 + a_{k+1,j}\delta_2 = v'_{k+1} < 0$$

$$a_{k+2,k}\delta_1 + a_{k+2,j}\delta_2 = v'_{k+2} > 0$$

This is always possible if the determinant of coefficients is different from zero. We now fix the value of j so that this condition is fulfilled. Such a j always exists, for suppose the contrary, namely that

(2) $\qquad a_{k+1,k} a_{k+2,j} - a_{k+2,k} a_{k+1,j} = 0 \qquad$ for all $\qquad j \neq k.$

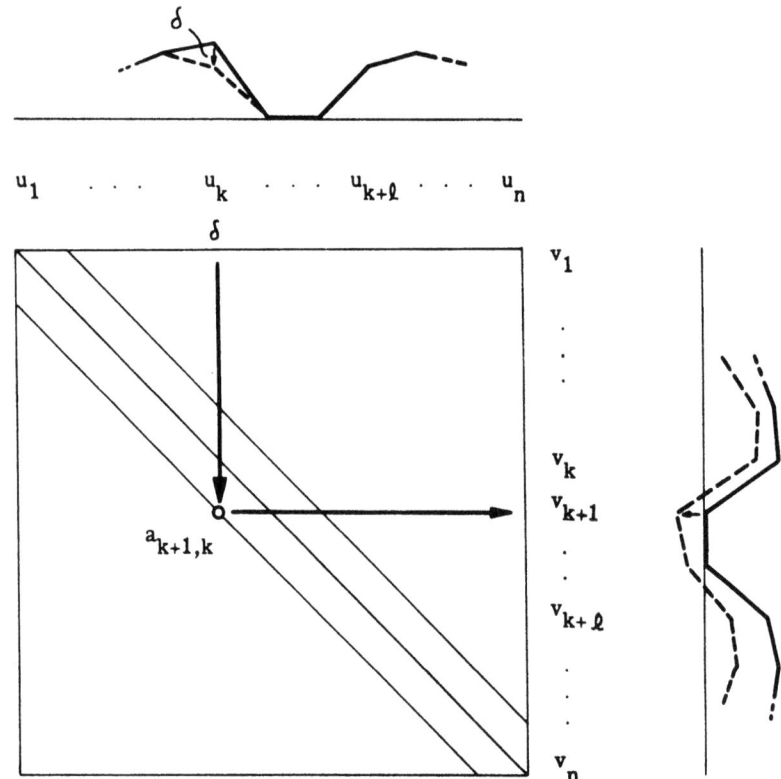

Figure 1. Schematic representation of the transformation $\mathcal{O} \tilde{\mathcal{M}} = \mathcal{A} \mathcal{O}$, showing the effect of the increment δ.

Taking into account the fact that the codiagonal element $a_{k+1,k}$ is different from zero, one can deduce from the equations (2) that $|\mathcal{O}l| = 0$. *) This, however, is impossible because of the assumption that $\mathcal{O}l$ is positive definite. Hence there is always at least one value of j for which (2) does not hold. For such a value of j the above inequalities are solvable.

If δ_1 and δ_2 are made sufficiently small, then

$$V(\Lambda\mathcal{O}') \geq V(\Lambda\mathcal{O}) + 2.$$

$|\delta_1|$ is to be chosen so small that $u_k u'_k > 0$. If $u_j \neq 0$ then similarly for δ_2. If $u_j = 0$, then since Case 1 is impossible, the nearest non-vanishing components to the right and left of u_j must have opposite signs. Hence the addition of δ_2 has no effect on the variation number.

Therefore

$$V(\tilde{\mathcal{M}}') = V(\tilde{\mathcal{M}}),$$

leading as before to the impossible relation

$$V(\Lambda\mathcal{O}') > V(\tilde{\mathcal{M}}').$$

Thus it is shown that both of the cases 1 and 2 are impossible.

There remains only one possibility:

If $u_i = 0$, then $u_{i-1} u_{i+1} < 0$. Q.E.D.

With the aid of this lemma Theorem 1 may be proved very easily.

The eigenvalues of (1) are positive because $\mathcal{O}l$ and $\mathcal{S}$ are positive definite. If an eigenvalue were multiple, then it would be possible to construct by linear combination a corresponding eigenvector with a vanishing first (or last) component. This would contradict Lemma 5.

---

*) If $a_{k+2,k} = 0$, then (2) implies that all other elements of the (k+2)-nd row are also zero. If $a_{k+2,k} \neq 0$, then (2) implies that the (k+1)-st and (k+2)-nd rows are proportional.

Therefore
$$\lambda_1 > \lambda_2 > \ldots > \lambda_n > 0.$$

Let $\breve{M}_k$ be the eigenvector which corresponds to λ_k.

If we set $\mathscr{L} = \mathcal{O}\mathcal{S}^{-1}$, *) then
$$\mathscr{L}\breve{M}_k = \lambda_k \breve{M}_k.$$

Let MO_0 be a linear combination of eigenvectors:
$$MO_0 = \sum_{i=r}^{s} c_i \breve{M}_i, \qquad 1 \leq r \leq s \leq n,$$

where $\qquad c_r \neq 0, \quad c_s \neq 0.$

Starting with MO_0 we iterate in both directions:
$$MO_\nu = \mathscr{L} MO_{\nu-1} \qquad\qquad MO_{-\nu} = \mathscr{L}^{-1} MO_{-\nu+1}$$

Since $\mathscr{L}$ is variation-diminishing and $\mathscr{L}^{-1}$ is variation-increasing, we have
$$V(MO_\nu) \leq V(MO_0) \leq V(MO_{-\nu})$$

If the MO_ν and $MO_{-\nu}$ are properly normalized after each step, then
$$\lim MO_\nu = \breve{M}_r \qquad\qquad \lim MO_{-\nu} = \breve{M}_s$$

By Lemmas 2 and 5 it follows that

(3) $\qquad\qquad V(\breve{M}_r) \leq V(MO_0) \leq V(\breve{M}_s).$

It remains to show that $V(\breve{M}_r) \neq V(\breve{M}_s)$ for $r \neq s$. Suppose the contrary:
$$V(\breve{M}_r) = V(\breve{M}_s) = V$$

---

*) $\mathscr{L}$ is also variation-diminishing.

Let $\mathcal{M}_0 = c_r \tilde{\mathcal{M}}_r + c_s \tilde{\mathcal{M}}_s$. By Lemma 4 it is possible to choose c_r and c_s so that $V(\mathcal{M}_0) \neq V$. This contradicts the relations (3).

It is thus shown that $\lambda_r > \lambda_s$ implies $V(\tilde{\mathcal{M}}_r) < V(\tilde{\mathcal{M}}_s)$.

Therefore
$$V(\tilde{\mathcal{M}}_k) = k - 1$$

Hence (3) may be written
$$r - 1 \leqslant V(\mathcal{M}) \leqslant s - 1. \qquad \text{Q.E.D.}$$

4. CHECKERBOARD TRANSFORMATION AND PROOF THAT THE COROLLARY IS A CONSEQUENCE OF THE THEOREM.

Let $\mathcal{O}^*$ be the matrix which is obtained from $\mathcal{O}$ by multiplying every even-numbered row and every even-numbered column of $\mathcal{O}$ by (-1). If all the elements of $\mathcal{O}$ are positive, then the signs of the elements of $\mathcal{O}^*$ are distributed checkerboard-fashion. For this reason the transformation $\mathcal{O} \rightarrow \mathcal{O}^*$ will be called the "checkerboard-transformation". *) Similarly, let $\mathcal{e}^*$ be the vector obtained from $\mathcal{e}$ by multiplying every even-numbered component of $\mathcal{e}$ by (-1)

(4) $\qquad$ If $\mathcal{O}\mathcal{e} = \mathcal{M}$ then $\mathcal{O}^* \mathcal{e}^* = \mathcal{M}^*$.

If $\mathcal{e}$ is an interior vector, then

(5) $\qquad V(\mathcal{e}^*) + V(\mathcal{e}) = n - 1.$

If $\mathcal{e}$ is a border vector, then

(6) $\qquad V(\mathcal{e}^*) + V(\mathcal{e}) < n - 1.$

---

*) A related transformation used in connection with the two dimensional Dirichlet problem is given by E.Stiefel, "Ueber einige Methoden der Relaxationsrechnung", Zeitschr. für angewandte Math. u. Phys., III (1952), p. 21.

(a) If $\mathcal{A}^*$ is variation-increasing, then $\mathcal{A}$ is variation-diminishing.*)

(b) The two problems

$$(1) \quad \mathcal{A}\wp = \lambda \vartheta \wp \quad \text{and} \quad (1^*) \quad \mathcal{A}^* \wp = \lambda \vartheta \wp$$

have the same eigenvalues. If $\tilde{m}_i$ are the eigenvectors of (1), then those of (1^*) are $\tilde{m}_i^*$.

(c) If $\mathcal{A}$ is symmetric and positive definite, then so is $\mathcal{A}^*$.

Statement (b) is a direct consequence of (4); statement (c) follows from (b). Statement (a) follows directly from (5) provided $\wp$ and $\textit{m} = \mathcal{A}\wp$ are both interior vectors. That (a) actually holds for all $\wp$ requires a little proof.

Lemma 6. If the relation $V(\mathcal{A}\wp) \leq V(\wp)$ holds for every interior vector $\wp$, then it holds for all $\wp$.

Proof: Suppose $\wp$ is a border vector. Let $\wp' = \wp + \delta_{\mathfrak{z}}$ be an <u>interior</u> vector such that

$$V(\wp') = V(\wp)$$

Let $m = \mathcal{A}\wp$ and $m' = \mathcal{A}\wp'$. Choose $|\delta|$ so small that

$$V(m') \geq V(m)$$

(See Lemma 1.) By hypothesis

$$V(m') \leq V(\wp')$$

Therefore $\quad\quad\quad\quad\quad\quad V(m) \leq V(\wp) \quad\quad\quad\quad\quad\quad$ Q.E.D.

---

*) If $\mathcal{A}$ is non-singular, the converse of (a) may be proved by applying (a) to the reciprocal matrix $\mathcal{A}^{-1}$.

Proof of Statement (a): Suppose that $\wp$ is an interior vector. By hypothesis

$$V(\mathcal{O}\!\ell^* \wp^*) \geq V(\wp^*) = n - 1 - V(\wp)$$

By (5) or (6): $\qquad V(\mathcal{O}\!\ell^* \wp^*) \leq n - 1 - V(\mathcal{O}\!\ell \wp)$

Therefore $\qquad V(\mathcal{O}\!\ell \wp) \leq V(\wp)$

for every interior vector $\wp$. By Lemma 6 this holds for all $\wp$.

It follows from the statements (a), (b) and (c) that the corollary is a consequence of the theorem. Suppose that the matrices $\mathcal{O}\!\ell^*$ and $\mathcal{S}$ of problem (1*) satisfy the hypotheses of the corollary. From (a) and (c) it follows that problem (1) satisfies the hypotheses of the <u>theorem</u>. From the theorem it follows that

$$V(\tilde{m}_k) = k - 1$$

where $\tilde{m}_k$ is the k-th eigenvector of problem (1). By (b) the eigenvectors of (1*) are $\tilde{m}_k^*$. Since the $\tilde{m}_k$ are interior vectors (Lemma 5), (4) is applicable:

$$V(\tilde{m}_k^*) = n - 1 - V(\tilde{m}_k) = n - k. \qquad \text{Q.E.D.}$$

The generalization concerning linear combinations of eigenvectors may be deduced as before by iterating in both directions.

Part II

5. MEAN VALUE OPERATOR.

It is evident that the property "variation-diminishing" is transitive.*)
Making use of this fact, we shall derive a rather general class of variation-diminishing linear transformations by means of an <u>iterative process.</u>
It will then be shown how the matrix of a transformation of this class may be calculated explicitly, and as a by-product of these considerations we shall obtain conditions under which the matrix is symmetric and positive definite.

Consider the index function y_i defined for the even integers $i = 2, 4, 6, \ldots$
In order to avoid fractional indices we shall reserve the odd integers $j = 1, 3, 5, \ldots$ for the intermediate points.

The variation number of the sequence

$$y_2, y_4, \ldots, y_{2n}$$

is always greater than or equal to that of the sequence of mean values:

$$y_2, \tfrac{1}{2}(y_2 + y_4), \tfrac{1}{2}(y_4 + y_6), \ldots, \tfrac{1}{2}(y_{2n-2} + y_{2n}), y_{2n}.$$

---

*) i.e. a transformation which consists in the successive application of several variation-diminishing transformations is also variation-diminishing. In other words, the product of several variation-diminishing matrices is also variation-diminishing.

Proof. The sequence

$$y_2, \; 1/2(y_2 + y_4), \; y_4, \; 1/2(y_4 + y_6), \; y_6, \; \ldots, \; y_{2n}$$

has exactly the same variation number as $y_2, y_4, \ldots, y_{2n}$. (Graphical representation!) The omission of the terms $y_4, y_6, \ldots, y_{2n-2}$ may lower but cannot raise the variation number. *)

Consider now the iterative process

(7)
$$\begin{aligned} m_j^{(1)} &= a_j^{(1)}(y_{j+1} + y_{j-1}) & a_j^{(1)} &> 0 \\ m_i^{(2)} &= a_i^{(2)}(m_{i+1}^{(1)} + m_{i-1}^{(1)}) & a_i^{(2)} &> 0 \\ m_j^{(3)} &= a_j^{(3)}(m_{j+1}^{(2)} + m_{j-1}^{(2)}) & a_j^{(3)} &> 0 \\ &\cdots \\ m_i^{(\nu)} &= a_i^{(\nu)}(m_{i+1}^{(\nu-1)} + m_{i-1}^{(\nu-1)}) & a_i^{(\nu)} &> 0 \end{aligned}$$

j = odd integers, i = even integers

where ν is even. If we set $y_i = 0$ for all i outside of i = 2, 4, ..., 2n, then the process (7) defines a sequence

(8) $$m_{-\nu+2}^{(\nu)}, \; m_{-\nu+4}^{(\nu)}, \; \ldots, \; m_{2n+\nu}^{(\nu)}$$

of $n + \nu$ terms whose variation number is less than or equal to that of the starting sequence $y_2, y_4, \ldots, y_{2n}$. The extra outside terms $(m_{-\nu+2}^{(\nu)}, \ldots, m_0^{(\nu)})$ and $(m_{2n+2}^{(\nu)}, \ldots, m_{2n+\nu}^{(\nu)})$ may be dropped from (8) without increasing its variation number. Therefore the matrix of the transformation

$$(y_2, \ldots, y_{2n}) \longrightarrow (m_2^{(\nu)}, \ldots, m_{2n}^{(\nu)})$$

is variation-diminishing. For convenience let us denote this transformation by

(9) $$\mathfrak{M} y = m$$

---

*) For various facts concerning the variation numbers of sequences see ref. 5.

This may also be written in the form

(9')
$$\mathcal{M}'\mathcal{H} = \mathcal{D}'m = m'$$

where $\mathcal{D}'$ is the diagonal matrix with the diagonal elements

$$\frac{1}{a_i^{(\nu)}}$$

In the special case $\nu = 2$, the matrix $\mathcal{M}'$ is

$$\mathcal{M}' = \mathcal{M}'_2 = \begin{bmatrix} (a_1 + a_3) & a_3 & & & \\ a_3 & (a_3 + a_5) & a_5 & & \\ & a_5 & (a_5 + a_7) & \cdots & \\ & & \cdots \cdots \cdots & & \\ & & & \cdots & (a_{2n-1} + a_{2n+1}) \end{bmatrix}$$

where for $a_j^{(1)}$ we have written simply a_j. $\mathcal{M}'_2$ is symmetric and its quadratic form may be written

$$a_1 x_2^2 + a_3 (x_2 + x_4)^2 + a_5 (x_4 + x_6)^2 + \ldots + a_{2n+1} x_{2n}^2$$

whence it is evident that $\mathcal{M}'_2$ is positive definite as well. All of the hypotheses of the theorem are fulfilled; hence the eigensolution of $\mathcal{M}'_2 \mathcal{H} = \lambda \mathcal{D}'_2 \mathcal{H}$ has all the properties described in Part I.

In the general case $\mathcal{M}'$ is symmetric and positive definite if the sequence

$$a_j^{(1)}, a_j^{(2)}, \ldots, a_j^{(\nu-2)}, a_j^{(\nu-1)}$$

is symmetric with respect to its middle term, in other words, if

(10)
$$\begin{aligned} a_j^{(k)} &= a_j^{(\nu-k)} & k &= 1, 3, 5, \ldots \\ a_i^{(\ell)} &= a_i^{(\nu-\ell)} & \ell &= 2, 4, 6, \ldots \end{aligned}$$

for all i and j.

Figure 2. Schematic representation of the transformation

$$\mathcal{M}' \mathcal{H} \mathbf{y} = \mathbf{m}'$$

or the quadratic form

$$Q = (\mathcal{H}\mathbf{y}, \mathcal{M}'\mathcal{H}\mathbf{y}).$$

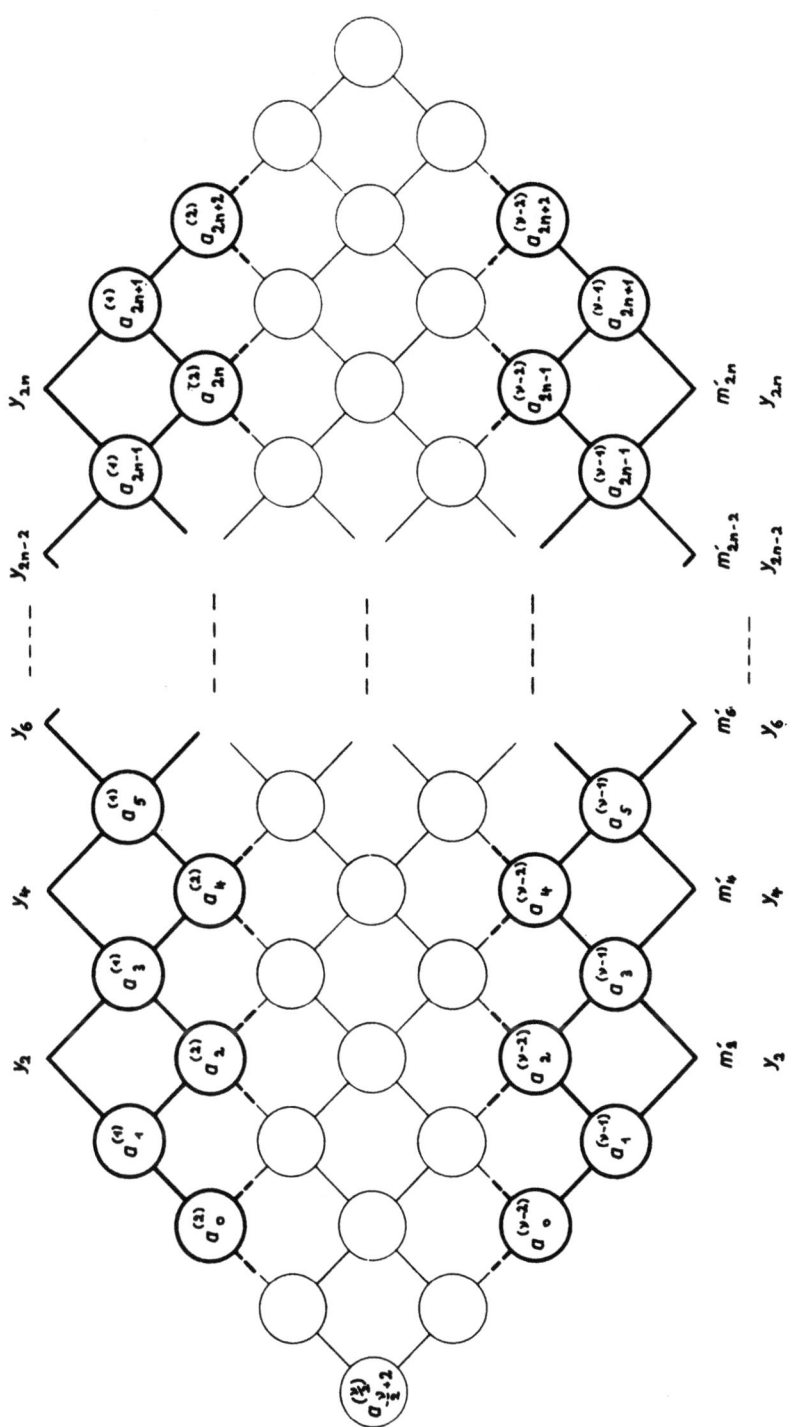

This may be shown the aid of Figure 2, which shows the structure of the matrix $\overline{\mathcal{M}}'$ schematically. If the conditions (10) are satisfied, then this scheme is symmetric with respect to its middle row. With every descending path through the scheme, i.e. every path which passes through each row exactly once, we associate a quantity p, where p is the __product__ of the $\nu - 1$ coefficients $a_s^{(r)}$ through which the path in question passes. The element $M'_{\alpha\beta}$ which lies in the α-th row and β-th column of $\overline{\mathcal{M}}'$ is equal to the __sum__ of the path products p for all paths which lead from $y_{2\beta}$ at the top of the scheme to $m'_{2\alpha}$ at the bottom. This means, in other words, all paths which lie within the rectangle whose upper corner is $y_{2\beta}$ and whose lower corner is $m'_{2\alpha}$. (See Figure 3a) That this is true can be seen by considering the transformation (9') for $\nu = 2$, then for $\nu = 4, \ldots$ etc.

Because of the symmetry of the scheme (conditions (10)), it is clear that $M'_{\alpha\beta} = M'_{\beta\alpha}$. __The symmetry of the scheme with respect to its middle row implies the symmetry of the matrix obtained from it.__

The quadratic form $Q = (\mathcal{M}\chi, \overline{\mathcal{M}}'\mathcal{M}\chi)$ is equal to the sum of all the path products p_h^* for all paths in the entire scheme where the end-points $y_{2\beta}$ and $y_{2\alpha}$ are included in each path:

$$Q = \sum_h p_h^*$$

We now collect terms in this sum with respect to the elements $a_k^{(\nu/2)}$ of the middle row. The set of all products p_h^* which contain the factor $a_k^{(\nu/2)}$ corresponds to the set of all paths which pass through $a_k^{(\nu/2)}$. This means, in other words, all paths which lie within the double triangle shown in Figure 3b.

Let the sum of all terms containing the factor $a_k^{(\nu/2)}$ be

$$a_k^{(\nu/2)} S_k.$$

S_k can be factored into the sum s_{k1} of all half-paths leading from the top of the scheme to $a_k^{(\nu/2)}$ on the one hand, and the sum s_{k2} of all half-paths leading from $a_k^{(\nu/2)}$ to the bottom of the scheme on the other hand:

$$S_k = s_{k1} s_{k2}.$$

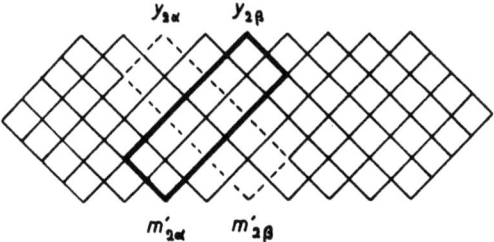

Fig. 3a. The element of the α-th row and β-th column equals the sum of the path products for all paths which lie within the heavily outlined rectangle.

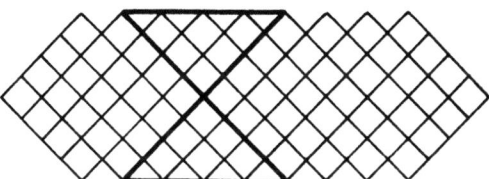

Fig. 3b. The set of all path products which contain the factor $a_k^{(\nu/2)}$ corresponds to the set of all paths which lie within the double triangle whose vertex is $a_k^{(\nu/2)}$.

Because of the symmetry of the scheme, $s_{k1} = s_{k2}$, hence S_k is a perfect square. Therefore Q can be written as a sum of perfect squares:

$$Q = \sum_k a_k^{(\nu/2)} s_k^2 \qquad S_k = S_{k1} = S_{k2}$$

with the positive coefficients $a_k^{(\nu/2)}$, whence it follows that $\mathcal{M}'$ is at least positive <u>semi</u>definite. It can be <u>strictly</u> semidefinite only if the system of equations $s_k = 0$ has a non-trivial solution $(y_2, y_4, \ldots, y_{2n})$. This, however, is impossible, for the first s_k on the left, namely $s_{-\nu/2 + 2}$ contains only y_2, whence

$$s_{-\nu/2 + 2} = 0 \quad \text{implies} \quad y_2 = 0.$$

The next one, namely $s_{-\nu/2 + 4}$, contains only y_2 and y_4 implying $y_4 = 0$, etc. Hence the system $s_k = 0$ has only the trivial solution. Therefore $\mathcal{M}'$ is <u>strictly</u> positive definite.

The codiagonal elements of $\mathcal{M}'$ are different from zero because of the assumption that the coefficients $a_s^{(r)}$ are strictly positive.

It is thus shown that the eigenvalue problem

$$\mathcal{M}'\eta = \lambda \mathcal{J}'\eta$$

(which is equivalent to $\mathcal{M}(\eta) = \lambda \mathcal{N}(\eta)$) fulfills all the hypotheses of the theorem for all (even) values of ν provided the conditions (10) are satisfied. From the theorem it follows that this problem has an inversely oscillatory eigensolution.

6. DIFFERENCE OPERATOR.

Replace the sums in (7) by differences:

(11)
$$d_j^{(1)} = a_j^{(1)}(y_{j+1} - y_{j-1}) \qquad a_j^{(1)} > 0$$
$$d_i^{(2)} = a_i^{(2)}(d_{i+1}^{(1)} - d_{i-1}^{(1)}) \qquad a_i^{(2)} > 0$$
$$\cdots\cdots\cdots\cdots \qquad \cdots\cdots$$
$$d_i^{(\nu)} = a_i^{(\nu)}(d_{i+1}^{(\nu-1)} - d_{i-1}^{(\nu-1)}) \qquad a_i^{(\nu)} > 0$$

j = odd integers, i = even integers.

If $\mathcal{L}$ is the matrix which is constructed from the equations (11) exactly as $\mathcal{M}$ was constructed from the equations (7), then

$$\mathcal{L} = \mathcal{M}^*,$$

where the asterisk refers to the checkerboard transformation. (See paragraph 4) If the conditions (10) are satisfied, then $\mathcal{L}$ fulfills the requirements of the corollary, whence it follows that $\mathcal{L}$ has an oscillatory eigensolution.

In paragraph 5 we prescribed "boundary conditions" for the mean value operator simply by dropping outside terms from the sequence (8). We shall now investigate some other boundary conditions for a special case of the difference operator (11). It will, however, be clear from this example that similar considerations hold for the general case as well.

The special case we wish to consider is the case

(12) $\qquad\qquad \nu = 4, \quad a_j^{(1)} = a_j^{(3)} = 1.$

The difference equation

(13) $\qquad\qquad d_i^{(4)} = \lambda y_i$

containing this operator corresponds to the differential equation

(14) $\qquad\qquad (py'')'' = \lambda q y \qquad p(x) > 0, \quad q(x) > 0,$

which occurs in the eigenvalue problem of the transversely vibrating rod with variable density and stiffness. We shall first summarize the known oscillatory properties of the eigensolution of the vibrating rod problem under various boundary conditions, and then show that the corresponding difference problems have analogous oscillatory properties by virtue of the corollary of Part. I.

We consider boundary conditions of the following type:

(15) $\qquad y_\alpha(a) = y_\beta(a) = 0, \qquad y_\gamma(b) = y_\delta(b) = 0$

where

$$y_0 = y$$
$$y_1 = y'$$
$$y_2 = py''$$
$$y_3 = (py'')'$$

For the sake of brevity the boundary conditions (15) will be designated by

$$(\alpha,\beta),(\gamma,\delta).$$

Of the six possible combinations (α,β), four are "symmetric": *)

| (0,1) | (0,2) | (1,3) | (2,3) |
| clamped | hinged | sliding | free |

and two are "asymmetric":

$\qquad\qquad\qquad$ (1,2) $\qquad$ (0,3).

If both (α,β) and (γ,δ) are symmetric, then the eigenvalue problem (14), (15) is self-adjoint and positive (semi-)definite. The eigenvalues are positive and simple with the possible exception of the first two, one or both of which

---

*) The justification of the term "symmetric" will become clear later.

may be equal to zero.

$$0 \underset{(=)}{\leq} \lambda_0 \underset{(=)}{\leq} \lambda_1 < \lambda_2 < \ldots < \lambda_k < \ldots$$

This information is tabulated below for all ten sets of self-adjoint boundary conditions. (No distinction is made between (α,β), $(\bar{\gamma},\bar{\delta})$ and $(\bar{\gamma},\bar{\delta})$, (α,β).)

i	(0,1), (0,1)	pos. def.
ii	(0,1), (0,2)	"
iii	(0,1), (1,3)	"
iv	(0,1), (2,3)	"
v	(0,2), (0,2)	"
vi	(0,2), (1,3)	"
vii	(0,2), (2,3)	$\lambda_0 = 0$
viii	(1,3), (1,3)	$\lambda_0 = 0$
ix	(1,3), (2,3)	$\lambda_0 = 0$
x	(2,3), (2,3)	$\lambda_0 = \lambda_1 = 0$

<u>In every case the eigenfunction which corresponds to the eigenvalue λ_k has exactly k simple roots in the interior of the interval (a,b).</u> 6), 7), 8).

If y is an eigenfunction of (14), (15), then the sequence $y_0, \ldots, y_3, \ldots$ is cyclical:

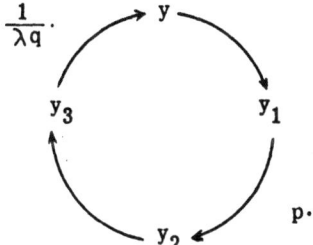

Each function in the cycle is the derivative of the preceding, multiplied every other time by a positive coefficient function. If the number of zeros in (a,b) of any of the four functions is given, then that of each of the other three is also determined by virtue of Rolle's theorem, which must hold for every pair of consecutive functions in the cycle. This is illustrated by the graphs of Figure 4.

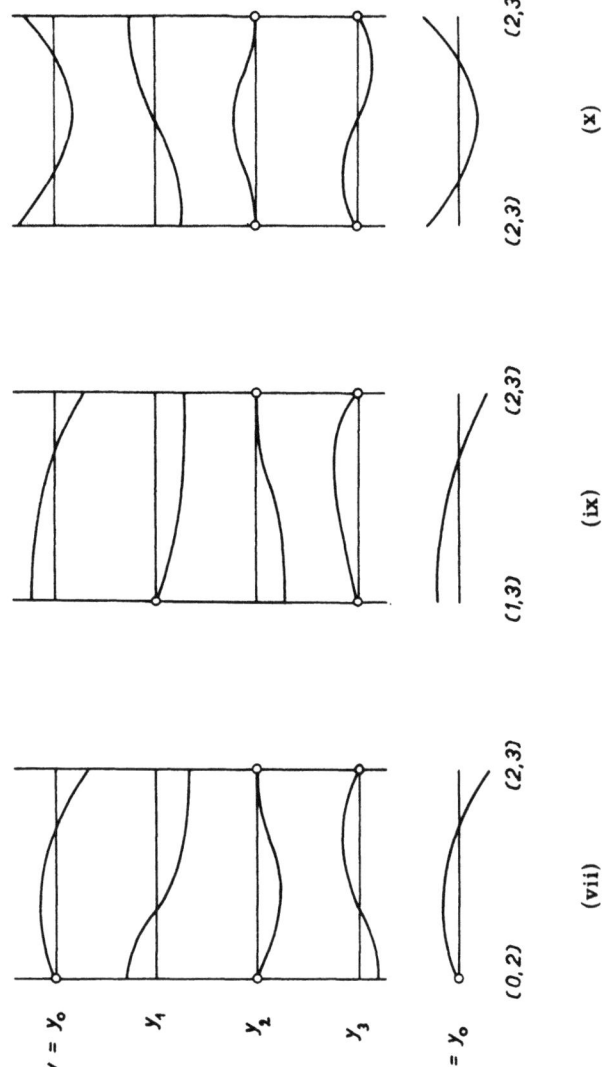

Fig. 4. Graphs illustrating the application of Rolle's theorem to the cyclical sequence $y_0, \ldots, y_3, \ldots$.

One can also deduce with the aid of Rolle's theorem several other interesting facts concerning problem (14), (15), e.g. the following:

No eigenfunction can have a multiple root in the interior of (a,b).

There is never more than one maximum or minimum between two roots of an eigenfunction.

None of the functions $y_0, \ldots, y_3$ has a zero at a or b which is not prescribed by the boundary conditions.

The operator $\frac{1}{q}(py'')''$ is variation-increasing with respect to the set of all sufficiently differentiable functions which satisfy the boundary conditions ("Vergleichsfunktionen").

By starting at the bottom of the cycle instead of at the top, one may rewrite equations (14) in terms of y_2:

(14') $$(\frac{1}{q} y_2'')'' = \lambda \frac{1}{p} y_2$$

The four symmetric boundary conditions carry over accordingly:

(See Fig. 4.)

y		y_2
(0,1)	$\longrightarrow$	(2,3)
(0,2)	$\longrightarrow$	(0,2)
(1,3)	$\longrightarrow$	(1,3)
(2,3)	$\longrightarrow$	(0,1)

By this means problems vii, ix and x may be written as strictly positive definite problems. The eigenfunctions which correspond to $\lambda_0 (= \lambda_1) = 0$ degenerate into the trivial solution of (14'), namely $y_2 = 0$, ($\lambda = 0$). Therefore all self-adjoint problems (14), (15) with the exception of (viii) may be treated as strictly positive definite. This fact will be useful in applying the corollary to the difference problems, as this was proved only for strictly positive definite matrices.

We now return to the difference problem. For convenience we write the operator (11) in the form of a difference table:

Y	: y_2	y_4	y_6	.....		y_{2n}
$D^{(1)}$:	$d_3^{(1)}$	$d_5^{(1)}$	$d_7^{(1)}$	.....		$d_{2n-1}^{(1)}$
$D^{(2)}$:		$d_4^{(2)}$	$d_6^{(2)}$	.....		$d_{2n-2}^{(2)}$
$D^{(3)}$:			$d_5^{(3)}$	$d_7^{(3)}$	.....	$d_{2n-3}^{(3)}$
$D^{(4)}$:				$d_6^{(4)}$	.....	$d_{2n-4}^{(4)}$

Starting from the vector $\mathcal{Y} = (y_2, y_4, \ldots, y_{2n})$ we wish to construct the transformed vector $\mathcal{D}^{(4)} = (d_2^{(4)}, d_4^{(4)}, \ldots, d_{2n}^{(4)})$.

Let the matrix of this transformation be $\mathcal{L}$:

$$\mathcal{L} \mathcal{Y} = \mathcal{D}^{(4)}$$

The components $(d_2^{(4)}, d_4^{(4)})$ and $(d_{2n-2}^{(4)}, d_{2n}^{(4)})$, however, are missing from the difference table. In order to determine these it is necessary to take the boundary conditions into account. The boundary conditions, therefore, are embodied in the matrix $\mathcal{L}$. We have now to determine whether $\mathcal{L}$ is variation-increasing under the boundary conditions in question. This may be done with the aid of the following lemma.

Lemma 7. *) If $V(X)$ is the variation number of the sequence
$X = x_1, x_2, \ldots, x_n$ and $V(D)$ the variation number of the sequence
$D = (x_2 - x_1), (x_3 - x_2), \ldots, (x_n - x_{n-1})$,

---

*) This lemma is an analogue of Rolle's theorem and is related to that upon which the mean value operator is based (paragraph 5) via the checkerboard transformation (paragraph 4). For proof see ref. 5.

then
$$V(D) \geq V(X) - 1.$$

If $x_1 = 0$ or $x_n = 0$, then
$$V(D) \geq V(X)$$

and if $x_1 = x_n = 0$, then
$$V(D) \geq V(X) + 1 \qquad (X \not\equiv 0)$$

Illustrative example. Consider the boundary conditions (v): (0,2), (0,2). We augment the difference table by setting

$y_0 = 0$ (thus determining $d_1^{(1)}, d_2^{(2)}, d_3^{(3)}, d_4^{(4)}$)

$d_0^{(2)} = 0$ (thus determining $d_1^{(3)}, d_2^{(4)}$)

The other end is treated similarly. For the thusly augmented difference table Lemma 7, applied successively to Y, $D^{(1)}$, ... yields

$$V(D^{(4)}) \geq V(Y).$$

Hence $\mathcal{L}$ is variation-increasing.

All boundary conditions of type (15) can be handled in a similar fashion. It is clear that $\mathcal{L}$ is in all cases variation-increasing. It remains to show that $\mathcal{L}$ is symmetric and positive definite.

If the elements of the augmented difference table which have been set equal to zero are dropped, then the end of the table assumes for each of the six boundary conditions (α,β) the characteristic shape shown in Figure 5. If we replace the elements $d_s^{(r)}$ of the difference table by the coefficients $a_s^{(r)}$ of the equations (11), then we obtain a scheme like that of Figure 2. The figures of Figure 5 may also be looked upon as schemes of this type. As was noted in connection with Figure 2, the matrix obtained from the scheme is symmetric if the scheme itself is symmetrical with respect to its midle row. Note that the four combinations (α,β) which were designated as "symmetric" actually correspond to symmetrical figures. Thus it is clear that the matrix $\mathcal{L}$ is symmetric if both (α,β) and (δ,δ) are "symmetric", i.e. if the boundary conditions (15) are self-adjoint.

Symmetric:

clamped   (0,1)

hinged   (0,2)

sliding   (1,3)

free   (2,3)

Asymmetric:

(1,2)

(0,3)

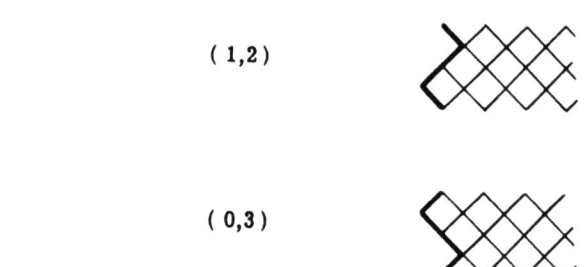

Figure 5

As for the definitenes of $\mathcal{L}$, it can be shown by considerations similar to those of paragraph 5 that $\mathcal{L}$ is strictly positive definite for the cases (i) through (vi). *) It has been shown that each of the cases (vii), (ix) and (x) can be reduced to one of the strictly positive definite cases.

The fact that the codiagonal elements of $\mathcal{L}$ are different from zero can be seen by applying the rule illustrated in Figure 3a to the schemes of Figure 5.

It is thus shown that the corollary of Part I is applicable in all cases with the possible exception of (viii).

The fact that the number of nodes is given by the variation number depends on the fact that two natural nodes never coincide. Two nodes may, however, be arbitrarily close together, but however close together they are, we have seen that they always appear in the solution of the difference problem.

The n - 1 sign variations of the n-th eigenvector of the difference problem are necessarily evenly distrubuted, i.e. one between each pair of consecutive terms of (y_i), no matter where the nodes of the <u>exact</u> eigenfunction may be. This means that under certain circumstances the approximation of the higher eigenfunctions is necessarily bad, but however bad it is, the number of nodes is always correct!

It would perhaps be interesting to investigate the situation with regard to other methods of approximation, e.g. the Ritz method, where the eigenfunctions are approximated by finite series of given functions.

7. EXTENSIONS.

Consider the general fourth order case where the coefficients $a_j^{(1)}$ and $a_j^{(3)}$ are subject to the conditions (10) i.e.

$$(16) \qquad a_j^{(1)} = a_j^{(3)} > 0$$

---

*) In cases (v) and (vi) split the diagram into its upper and lower halves. Each of these halves is symmetrical. Each half represents the system of equations $s_k = 0$. Apply the method of paragraph 5 to this system.

but are otherwise arbitrary. In this case the corresponding differential operator is

(17)
$$\frac{1}{q}\left\{r\left[p(ry')'\right]'\right\}'$$

$$p(x) > 0, \quad r(x) > 0, \quad q(x) > 0$$

(17) is self-adjoint. The general self-adjoint differential operator of the fourth order can be written in the form

(18)
$$\frac{1}{q}\left[(P^2 y'')'' + (Ry')'\right] \qquad P^2(x) > 0.$$

If $P(x)$ and $R(x)$ are of such a nature that the linear, second order differential equation

(19)
$$P^2 u'' + (R - PP'')u = 0$$

has a positive solution $u(x) > 0$ for $a \leqslant x \leqslant b$, then (18) is reducible to the form (17) for $a \leqslant x \leqslant b$. This transformation is given by*)

$$p = u^2, \qquad r = \frac{P}{u}$$

The corollary of Part I, therefore, is also applicable to the difference operator which corresponds to the more general differential operator (18) (under the condition that (19) has a positive solution). There are a number of phenomena of elastomechanics which are governed by equations of the form (18). For some examples see Table I in the back of Collatz's "Eigenwertaufgaben".[10] (p. 412) The condition (19) can, for example, always be satisfied if P is a constant and R is negative. This is actually the case for most of the examples given by Collatz.

The considerations of paragraph 6 could evidently be extended to difference operators of the form (11) of any (even) order.

---

*) This may be verified by differentiating out (17) and (18) and comparing terms. As a check, note the case $R \equiv 0$.

8. APPLICATIONS.

In calculating an eigenvalue numerically, it is desirable to find definite upper and lower bounds for it rather than only an approximate value. There are various methods for calculating such bounds, most of which operate somewhat as follows: A formula or process is given which when applied to <u>any</u> trial vector yields two numbers which enclose at least one eigenvalue between them. The interval between the numbers comes out larger or smaller depending on how close the trial vector happens to be to an eigenvector. The trick, of course, is to choose the trial vector so that the interval is as small as possible. Supposing this is done, one still does not know <u>which</u> eigenvalue (according to its relative position in the spectrum) has been enclosed. This information can be very important. In the problem of finding the critical speeds of a rotating shaft, for example, the neighborhoods of the eigenvalues are danger regions. Here, it is not so important to know where the eigenvalues are as to know where they aren't! It is therefore important to know which ones have been found and which, if any, have been missed.

If the analytic problem being investigated is known to satisfy an oscillation theorem, one generally assumes in practice that the enclosed eigenvalue is the k-th if the trial function (vector) has k - 1 nodes. This assumption is not entirely watertight for two reasons: first, because the oscillation theorem may no longer be valid for the approximate problem, and second, because the trial function (vector) is itself only approximate.

In connection with the method of L. Collatz, however, this assumption is correct provided both bounds are positive. Modified so as to take the oscillation theorem of Part I into account, Collatz's theorem may be stated as follows. [9]) (See also [10]), p. 289.)

Theorem 2. Suppose that the matrices $\mathcal{O}$ and $\mathcal{B}$ of the eigenvalue problem

(1) $$\mathcal{O} e = \lambda \mathcal{B} e$$

satisfy the hypotheses of Theorem 1. Let the trial vector

$$\tilde{M} = (u_1, \ldots, u_n)$$

have the variation number $V(\tilde{m}) = k - 1$. From $\tilde{m}$ and the transformed vector

$$\mathcal{A}\mathcal{O} = \mathcal{A}\tilde{m} = (v_1, \ldots, v_n)$$

we form the quotients

$$q_i = \frac{v_i}{u_i d_i}$$

where the d_i (>0) are the diagonal elements of $\mathcal{D}$. If all of the q_i are positive, then the k-th eigenvalue in the sequence

$$\lambda_1 > \lambda_2 > \ldots > \lambda_n > 0$$

lies between Max(q_i) and Min(q_i).

If $\mathcal{A}$ satisfies the hypotheses of the corollary, then the $(n - k + 1)$-st eigenvalue lies between Max(q_i) and Min(q_i).

Proof. Let $\mathcal{Q}$ be the (positive) diagonal matrix whose diagonal elements are the quotients q_i. The trial vector $\tilde{m}$ satisfies the equation

$$\mathcal{A}\tilde{m} = \mathcal{A}\mathcal{O} = \mathcal{Q}\mathcal{D}\tilde{m}$$

According to the theorem of Part I the problem

(1') $$\mathcal{A}\boldsymbol{\gamma} = \lambda^* \mathcal{Q}\mathcal{D}\boldsymbol{\gamma}$$

has an inversely oscillatory eigensolution (as well as problem (1)). Since $V(\tilde{m}) = k - 1$, $\tilde{m}$ must be the k-th eigenvector of (1'), the corresponding eigenvalue being equal to one: $\lambda_k^* = 1$. We now compare the following three eigenvalue problems:

$$\mathcal{A}\boldsymbol{\gamma} = \lambda q_{max} \mathcal{D}\boldsymbol{\gamma}$$
$$\mathcal{A}\boldsymbol{\gamma} = \lambda^* \mathcal{Q}\mathcal{D}\boldsymbol{\gamma}$$
$$\mathcal{A}\boldsymbol{\gamma} = \bar{\lambda} q_{min} \mathcal{D}\boldsymbol{\gamma}$$

From the Maximum-minimum Principle of Courant*), it follows that

$$\underline{\lambda}_i \leq \lambda_i^* \leq \overline{\lambda}_i \qquad i = 1, \ldots, n$$

In particular
$$\underline{\lambda}_k \leq 1 \leq \overline{\lambda}_k$$

Since
$$\lambda_i = \underline{\lambda}_i q_{max} = \overline{\lambda}_i q_{min}$$

it follows that
$$q_{max} \geq \lambda_k \geq q_{min} \qquad \text{Q.E.D.}$$

Exactly analogous theorems, modified with the aid of the classical, analytical oscillation theorems, hold for differential and integral eigenvalue problems. (See [10], p. 126.)

With the aid of Theorem 2 (or its analytic counterpart) bounds for any desired eigenvalue, identified by its index, may be calculated directly without previous information.

A process of "enlightened trial and error" by which these bounds may be improved is described by Zurmühl, [12].

---

*) See ref. 10, p. 289.

Appendix

9. VIBRATING ROD WITH INTERMEDIATE SUPPORTS.

Very often in the vibrating rod problem there are one or more supports or bearings <u>between</u> the end-points. These are usually idealized as sharp fulcra which completely obstruct transverse motion, but through which the bending moment is freely transmitted. In mathematical terms, the connection conditions at such support points are the following: (See equation (15) ff.)

(20)
$$y_0 = y = 0$$
$$y_1 = y' \quad \text{continuous}$$
$$y_2 = py'' \quad \text{continuous}$$
$$y_3 = (py'')' \quad \text{(in general) discontinuous}$$

The first two conditions are geometrically plausible. y_2 is essentially the bending moment and y_3 is essentially the transversal shearing force. The third condition expresses the fact that the fulcrum offers no resistance to moments, and the fourth condition expresses the fact that transverse reaction forces are (in general) present at the support points.

How are these conditions to be interpreted for the difference problem? If a function is discontinuous at a point, then the derivative is undetermined at that point. Accordingly, we assume for the difference problem that at a point of discontinuity the <u>difference</u> is undetermined. Applying this rule to the difference table (paragraph 6) we obtain the symmetrical scheme shown in Figure 6. The difference operator represented by Fig. 6 operates on the index function (vector) $(y_i) = (y_2, y_4, \ldots, y_{2n})$. Since the value of the function at the support point is fixed (= zero), no index is assigned to this point; it is simply passed over.

If the lattice of Fig. 6 is regarded as a scheme of the type shown in

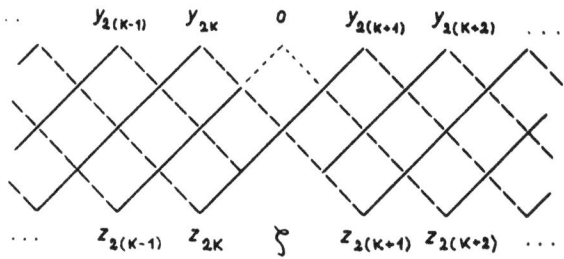

Fig. 6. Difference scheme in the neighborhood of a support point.

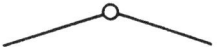
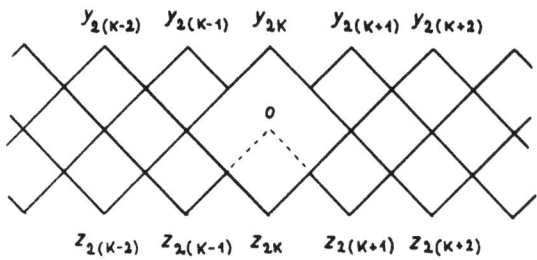

Fig. 7. Difference scheme in the neighborhood of a joint.

Fig. 2 (instead of as a difference table), then all of the considerations of paragraph 5 may be applied, provided proper attention is paid to the signs. In the mean value operator everything was positive; here, we are concerned essentially with a checkered distribution of signs.

With each path segment in Fig. 6 which descends toward the left (solid line) we associate the sign factor (+1) and with each path segment which descends toward the right (dotted line) we associate the sign factor (-1). Each path product p_h is to be multiplied by the product of the sign factors of all the segments in the path. Taking this rule of signs into account, we may construct the matrix of our difference operator from Fig. 6 according to the rule illustrated in Fig. 3a (paragraph 5). Let us call this matrix $\mathcal{L}$.

The difference matrices of paragraph 6 were all obtainable from mean value matrices via the checkerboard transformation; the signs of the elements were distributed strictly checkerboard fashion. This is no longer the case. Because of the irregularities in the neighborhood of the support point, there are two elements of $\mathcal{L}$, namely $b_{k,k+1}$ (k-th row and (k+1)-st column) and $b_{k+1,k}$, whose signs violate the checkerboard rule. $\mathcal{L}$, therefore, is not variation-increasing. By means of a simple modification, however, $\mathcal{L}$ can be transformed into a matrix which is variation-increasing. The following general considerations will be of use in explaining this modification.

Definition. Let $\overline{\mathcal{A}}$ be the matrix which is obtained from $\mathcal{A}$ by multiplying all the rows and all the columns of $\mathcal{A}$ whose indices are j+1, ..., n by (-1). If, for example, all the elements of $\mathcal{A}$ are positive, then $\overline{\mathcal{A}}$ is divided into four rectangular fields within each of which the signs are uniform. The signs of the fields are distributed checkerboard-fashion as indicated below.

```
          1 ... j j+1 ... n
      1 ┌───────┬───────┐
      ⋮ │       │       │
        │   +   │   −   │
      j │       │       │
    j+1 ├───────┼───────┤
      ⋮ │       │       │
        │   −   │   +   │
      n │       │       │
        └───────┴───────┘
```

47

For this reason the transformation $\mathcal{O}\!\!\!l \longrightarrow \overline{\mathcal{O}\!\!\!l}$ will be called a <u>checkerboard transformation of order one</u>. *) The index j will be called the <u>transition point</u>. Similarly let $\overline{\eta}$ be the vector obtained from η by multiplying the components of η whose indices are j+1, ..., n by (-1).

$$\mathcal{O}\!\!\!l\,\eta = \mathcal{M}\gamma \quad \text{implies} \quad \overline{\mathcal{O}\!\!\!l}\,\overline{\eta} = \overline{\mathcal{M}\gamma}$$

Suppose that symmetric boundary conditions of the type (15) are prescribed on the ends, and that at some intermediate point connection conditions (20) are given. Construct the corresponding lattice (Figs. 5 and 6), and from the lattice construct the corresponding <u>mean value</u> matrix according to the rule of Fig. 3a. Let this matrix be denoted by $\mathcal{M}$.

$\mathcal{M}$ <u>is variation-diminishing, and provided the problem is positive definite, all other hypotheses of Theorem 1 are also satisfied.</u>**)

We have noted that the difference matrix $\mathcal{L}$ is not equal to $\mathcal{M}^*$ as was the case in paragraph 6. Instead, we have the relation

$$\overline{\mathcal{L}} = \mathcal{M}^*$$

---

*) $\mathcal{O}\!\!\!l \longrightarrow \mathcal{O}\!\!\!l^*$ is the checkerboard transformation of order n-1.

**) Imagine the scheme to be entirely filled-in in the neighborhood of the support point. (For the missing term between y_{2k} and $y_{2(k+1)}$ substitute the value zero.) By paragraph 5 the transformation represented by the filled-in scheme is variation-diminishing. The variation number of the sequence $(z_2, ..., z_{2k}, \zeta, z_{2(k+1)}, ..., z_{2n})$ is not increased if the undetermined element ζ is dropped. Therefore the transformation $\mathcal{M}\eta = \gamma$ is also variation-diminishing. That the other conditions of the theorem are fulfilled may be verified by the methods of paragraphs 5 and 6.

where the bar refers to the checkerboard transformation of order one whose transition point is k, i.e. whose transition point coincides with the support point. Since $\overline{\mathcal{M}}$ has an inversely oscillatory eigensolution, $\overline{\mathcal{M}}{}^*$ has an oscillatory one. (paragraph 4)

If the eigenvectors of $\mathcal{L}$ are $\check{\mathcal{M}}_i$, then the system $\overline{\check{\mathcal{M}}}_i$ is oscillatory.

In case there are m intermediate supports one has merely to apply the appropriate checkerboard transformation of order m. This is a plausible result. The checkerboard transformation "removes" the forced nodes at the support points, leaving only the natural nodes to be counted.

The cyclical transformation (14') transforms a support point into a joint as indicated in Fig. 7. With regard to the jointed rod the situation is simpler. In this case the corollary of Part I is directly applicable without modification.

These results may be summarized as follows. The eigenvalue problem of the inhomogeneous vibrating rod which is clamped, hinged, free or sliding on the ends and which has any number of joints in between can be approximated by a difference eigenvalue problem whose eigensolution is oscillatory provided the problem is positive definite, i.e. provided the boundary conditions are such that the rest position of the rod is mechanically fixed. If there are m intermediate <u>supports</u> then the eigensolution can be transformed into an oscillatory one by application of the checkerboard transformation of the m-th order whose m transition points coincide with the m support points. [11])

ZUSAMMENFASSUNG

Man betrachte die lineare Transformation

$$a_{11}x_1 + a_{12}x_2 + \ldots + a_{1n}x_n = y_1$$
$$a_{21}x_1 + a_{22}x_2 + \ldots + a_{2n}x_n = y_2$$
$$\vdots \qquad\qquad\qquad \vdots$$
$$a_{n1}x_1 + a_{n2}x_2 + \ldots + a_{nn}x_n = y_n$$

die bequem dargestellt werden kann durch die Symbole

$$A\boldsymbol{x} = \boldsymbol{y}.$$

Es sei die Anzahl der Zeichenwechsel in der Folge $\boldsymbol{x} = x_1, \ldots, x_n$ durch $V(\boldsymbol{x})$ bezeichnet.

Die Matrix A heisst <u>variationsvermindernd</u>, wenn für jedes n-tupel von Zahlen $\boldsymbol{x} = x_1, \ldots, x_n$ die Beziehung

$$V(\boldsymbol{y}) \leqslant V(\boldsymbol{x})$$

gilt. *)

Die vorliegende Arbeit befasst sich mit dem Eigenwertproblem

(1) $$A\boldsymbol{x} = \lambda D \boldsymbol{x},$$

wo A symmetrisch und positiv definit ist, und wo D eine positive Diagonalmatrix ist. **)

---

*) Gilt $V(\boldsymbol{y}) \geqslant V(\boldsymbol{x})$ für jedes $\boldsymbol{x}$, so heisst A <u>variationsvermehrend</u>.

**) d.h. $d_{ij} \begin{cases} = 0, & i \neq j \\ > 0, & i = j \end{cases}$

Ist A auch noch variationsvermindernd, und sind alle codiagonale Elemente $a_{i,i-1}$, $a_{i,i+1}$ verschieden von Null, dann besitzen die Eigenvektoren und Eigenwerte des Problems (1) eine Reihe interessanter und nützlicher Eigenschaften:

(a) Sämtliche Eigenwerte sind positiv und einfach:
$$\lambda_1 > \lambda_2 > \ldots > \lambda_n > 0$$

(b) *) Ist $\tilde{u}_k$ der dem Eigenwert λ_k entsprechende Eigenvektor, so gilt
$$V(\tilde{u}_k) = k - 1.$$

(c) **) Ist $\tilde{u} = (u_1, \ldots, u_n)$ ein Eigenvektor, so gilt für die Komponenten
$$u_1 \neq 0$$
$$u_i = 0 \quad \text{nur dann, wenn} \quad u_{i-1} u_{i+1} < 0$$
$$u_n \neq 0$$

Die Behauptungen (a) und (b) sind denjenigen des klassischen Sturm-Liouvilleschen Oszillationstheorems in ganz anschaulicher Weise analog.

<u>Prinzip des Beweises.</u> Wenn die Behauptung (c) einmal feststeht, so ist der Beweis von (a) und (b) im Prinzip einfach. Die Eigenwerte sind positiv wegen der positiven Definitheit, und keiner ist mehrfach, denn sonst könnte man durch lineare Kombination einen zugehörigen Eigenvektor mit verschwindender erster (oder letzter) Komponente bilden, was Behauptung (c) widersprechen würde.

---

*) Allgemeiner gilt folgendes. Ist $\tilde{u}_0$ irgendeine Kombination von Eigenvektoren:
$$\tilde{u}_0 = (c_r \tilde{u}_r + \ldots + c_s \tilde{u}_s) \qquad 1 \leq r \leq s \leq n$$
dann ist $r - 1 \leq V(\tilde{u}_0) \leq s - 1$.

**) Einen Vektor mit der Eigenschaft (c) nennen wir einen "inneren" Vektor aus folgendem Grunde. Es existiert im n-dimensionalen Raum dann und nur dann eine Umgebung des Vektors $\tilde{u}$, in der alle Vektoren dieselbe Wechselzahl wie $\tilde{u}$ selbst besitzen, wenn $\tilde{u}$ die Eigenschaft (c) hat. Nach (c) ist also jeder Eigenvektor ein innerer Vektor.

Um (b) zu beweisen, iteriert man nach beiden Richtungen. Ausgehend von einem Vektor m_0, dessen Entwicklung nach Eigenvektoren durch

$$m_0 = \sum_{i=r}^{s} c_i \tilde{m}_i \qquad 1 \leq r \leq s \leq n$$

gegeben ist, erhält man durch Iteration mit $B = AD^{-1}$ sowie mit B^{-1} zwei Folgen von Vektoren, die gegen $\tilde{m}_r$ bzw. $\tilde{m}_s$ konvergieren:

$$m_\nu = B m_{\nu-1} \qquad \lim m_\nu = \tilde{m}_r$$

$$m_{-\nu} = B^{-1} m_{-\nu+1} \qquad \lim m_{-\nu} = \tilde{m}_s$$

(Dabei müssen die m_ν nach jedem Schritt normiert werden.) Da aber B variationsvermindernd und B^{-1} variationsvermehrend ist, gilt

$$V(m_\nu) \leq V(m_{-\nu}).$$

Man kann zeigen, dass dies auch im Grenzfall gilt:

$$V(\tilde{m}_r) \leq V(\tilde{m}_s).$$

Indem noch gezeigt wird, dass $V(\tilde{m}_r) \neq V(\tilde{m}_s)$ für $r \neq s$, ist der Beweis erbracht.

Ist A <u>variationsvermehrend</u> und gelten auch die übrigen Voraussetzungen, dann gilt statt (b) die Behauptung

(b*) $\qquad V(\tilde{m}_k) = n - k.$

Dies lässt sich aus dem Hauptsatz mit Hilfe der sogenannten "Schachbrett-Transformation" herleiten.

Es stellt sich nun die Frage, ob die kritische Voraussetzung, (i.e. dass die Matrix A variationsvermindernd, bzw. -vermehrend sei) praktisch brauchbar ist. Im zweiten Teil der Arbeit wird eine Klasse derartiger Matrizen definiert, welche mehrere für die Praxis wichtige Spezialmatrizen umfasst. Am anschaulichsten (wenn auch etwas ungenau) kann diese Klasse vielleicht folgendermassen erklärt werden. Man betrachte den Differentialoperator

(2) $$z(x) = \frac{d}{dx}\left[p_{\mu-1}\frac{d}{dx}\left[\ldots\left[p_2\frac{d}{dx}\left[p_1\frac{dy}{dx}\right]\right]\ldots\right]\right]$$

wo μ eine gerade Zahl ist, und wo $p_1(x), \ldots, p_{\mu-1}(x)$ positive, genügend differenzierbare Funktionen sind. Ist ferner

$$p_i(x) = p_{\mu-i}(x) \qquad i = 1, 2, \ldots$$

so ist (2) selbstadjungiert. Abgesehen von diesen Einschränkungen dürfen die $p_i(x)$ beliebig gewählt werden. Man denke sich den Operator (2) in μ Etappen zerlegt:

$$z_1 = p_1 y'$$
$$z_2 = p_2 z_1'$$
$$\ldots\ldots\ldots$$
$$z = z'_{\mu-1}$$

An den Randpunkten $x = a$ und $x = b$ sei vorgeschrieben, dass von den μ Funktionen $y, z_1, \ldots, z_{\mu-1}$ genau die Hälfte verschwinden. Es folgt nun aus dem Rolleschen Satz, dass im Intervall (a, b) die Funktion $z(x)$ nie weniger Nullstellen besitzt als $y(x)$. Der Operator (2) ist also in diesem Sinne variationsvermehrend. Konstruiert man den Differenzenoperator, dem (2) entspricht (unter Berücksichtigung der oben erwähnten Randbedingung), so ist der resultierende Matrixoperator tatsächlich variationsvermehrend. Diese Matrix hängt von einer Anzahl beliebiger Grössen ab, die von den beliebigen Funktionen $p_i(x)$ herstammen.

Das Eigenwertproblem des transversal schwingenden, inhomogenen Stabes stellt wohl das wichtigste Anwendungsgebiet dieser Ideen dar. Dieses Problem wird daher im Detail besprochen, und insbesondere wird in einem Anhang der mehrfach gestützte, sowie der mehrfach gelenkig gegliederte Stab behandelt.

Der Satz ist besonders nützlich bei der Berechnung von oberen und unteren
Schranken für die Eigenwerte. Man betrachte wiederum das Eigenwertproblem (1). Von einem Versuchsvektor ρ ausgehend, bilde man die
Vektoren

$$(u_1, \ldots, u_n) = \tilde{w} = A\rho$$
$$(v_1, \ldots, v_n) = w = D\rho$$

und die Quotienten

$$q_i = \frac{u_i}{v_i}.$$

Nach dem Einschliessungssatz von L. COLLATZ liegt (mindestens) ein
Eigenwert zwischen den Schranken $\text{Max}(q_i)$ und $\text{Min}(q_i)$. Die Nummer
des eingeschränkten Eigenwerts, d.h. seine relative Lage im Spektrum,
ist noch unbekannt. Ist aber A variationsvermindernd (bzw. -vermehrend) und gelten auch die übrigen Voraussetzungen, so kann diese Auskunft ohne weiteres ermittelt werden. Man hat in diesem Falle nur die
Zeichenwechsel des Versuchsvektors ρ zu zählen. Ist $V(\rho) = k$,
so liegt der $(k+1)$-te (bzw. der $(n-k)$-te) Eigenwert zwischen den
Schranken. *)

---

*) Dies gilt mit Sicherheit nur dann, wenn $\text{Min}(q_i) > 0$, was allerdings
bei brauchbaren Schranken fast immer der Fall sein wird.

REFERENCES

[1]) Krejn, M., "Sur les fonctions de Green non-symétriques oscillatoires des operateurs différentiels ordinaires", C. R. Acad. Sci. URSS, N.s. 25, (1939), 643-646.

[2]) Gantmakher, F., et M. Krejn, "Sur les matrices complètement non-négatives et oscillatoires", Compositio Math. 4 (1937), 445-476.

[3]) Schoenberg, I. J., "Ueber variationsvermindernde lineare Transformationen", Math. Zeitschr., 32 (1930), p. 321.

[4]) Motzkin, T., "Lineare Ungleichungen", Diss., Basel (1933).

[5]) Polya, G., and G. Szegö, "Aufgaben und Lehrsätze aus der Analysis", (Springer, Berlin 1925), V. Abschn., Kap. 1, Aufg. 7, 13, 15.

[6]) Haupt, O. "Oszillationstheoreme", Diss., Würzburg (1911).

[7]) Janczewsky, S. A., "Oscillation theorems for the differential boundary value problems of the fourth order", Annals of Math., 2. ser., 29, (1928), pp. 521-542; Part II of same, ibid., 3. ser., 31 (1930), pp. 663-680.

[8]) Davis, Amer. Journ. of Math., 47 (1925)

[9]) Collatz, L., "Einschliessungssatz für die charakteristischen Zahlen von Matrizen", Math. Zeitschr. 48 (1942), pp. 221-226.

[10]) Collatz, L., "Eigenwertaufgaben mit technischen Anwendungen", Leipzig (1949) (Geest & Portig).

[11]) Hohenemser, K., u. W. Prager, "Ueber die Anzahl der Knotenpunkte bei erzwungenen u. freien Schwingungen, Zeitschr. angew. Math. u. Mech., 11 (1931).

[12]) Zurmühl, "Matrizen, eine Darstellung für Ingenieure", (Springer, Berlin, 1950)

Birkhäuser Verlag, Basel und Stuttgart

Mitteilungen
aus dem Institut für angewandte Mathematik
an der Eidgenössischen Technischen Hochschule Zürich

Herausgegeben von Prof. Dr. E. Stiefel

Nr. 1

Entwurf eines elektronischen Rechengerätes unter besonderer Berücksichtigung der Erfordernis eines minimalen Materialaufwandes bei gegebener mathematischer Leistungsfähigkeit. Von *Ambros P. Speiser* (2. Auflage 1954). 67 Seiten.
Vorstudien für die Entwicklung eines Rechenautomaten mittlerer Grösse, angepasst an die Bedürfnisse eines mathematischen Instituts. Grundlage für die elektronische Rechenmaschine ERMETH der Eidgenössischen Technischen Hochschule Zürich. Der Entwurf sieht 1000 Elektronenröhren und 300 elektromagnetische Relais vor; als Speicher dient eine magnetische Trommel.

Nr. 2

Programmgesteuerte digitale Rechengeräte (elektronische Rechenmaschinen). Von *Heinz Rutishauser, Ambros Speiser* und *Eduard Stiefel* (Nachdruck 1958). 102 Seiten.
Inhalt: Grundlagen und wissenschaftliche Bedeutung – Organisation und Arbeitsweise – Arithmetische Prinzipien – Vorbereitung von Rechenplänen – Physikalische Grundlagen.

Nr. 3

Automatische Rechenplanfertigung bei programmgesteuerten Rechenmaschinen. Von *Heinz Rutishauser* (Nachdruck 1961.) 45 Seiten.
Vorstudien für die Automation des Programmierens für elektronische Rechenautomaten.

Nr. 4

An Oscillation Theorem for Algebraic Eigenvalue Problems and its Applications. By *Frank W. Sinden.* (1954.) 57 pages.
In der Theorie der Eigenwertprobleme bei gewöhnlichen Differentialgleichungen (z. B. kritische Drehzahlen) spielen die sogenannten Oszillationstheoreme eine grosse Rolle, welche Aufschluss geben über die Anzahl der Knoten der Oberschwingungen. Der Verfasser überträgt diese Theorie auf die Differenzenrechnung, womit Oszillations-Theoreme auch in der numerischen Mathematik ihre Anwendung finden können.

Zu beziehen durch Ihre Buchhandlung – Obtainable from your bookseller
Commandes à votre libraire

MIX
Papier aus verantwortungsvollen Quellen
Paper from responsible sources
FSC® C105338

If you have any concerns about our products,
you can contact us on
ProductSafety@springernature.com

In case Publisher is established outside the EU,
the EU authorized representative is:
**Springer Nature Customer Service Center GmbH
Europaplatz 3, 69115 Heidelberg, Germany**

Printed by Libri Plureos GmbH
in Hamburg, Germany